Esteban Favela-Chávez
Alejandra C. Favela-Gaytán

Description and Management of Ornamental Trees for Use in Urban Areas

Esteban Favela-Chávez
Alejandra C. Favela-Gaytán

Description and Management of Ornamental Trees for Use in Urban Areas

ScienciaScripts

Imprint
Any brand names and product names mentioned in this book are subject to trademark, brand or patent protection and are trademarks or registered trademarks of their respective holders. The use of brand names, product names, common names, trade names, product descriptions etc. even without a particular marking in this work is in no way to be construed to mean that such names may be regarded as unrestricted in respect of trademark and brand protection legislation and could thus be used by anyone.

Cover image: www.ingimage.com

This book is a translation from the original published under ISBN 978-620-2-15069-9.

Publisher:
Sciencia Scripts
is a trademark of
Dodo Books Indian Ocean Ltd. and OmniScriptum S.R.L publishing group

120 High Road, East Finchley, London, N2 9ED, United Kingdom
Str. Armeneasca 28/1, office 1, Chisinau MD-2012, Republic of Moldova, Europe
Printed at: see last page
ISBN: 978-620-7-63579-5

Contents

CHAPTER I

I. Introduction.

Ornamental species, especially native ones, have adapted perfectly to field conditions for use in urban areas, this is facilitated by the fact that these species have certain tolerances to drought, high and low temperatures and most of them adapt to any type of soil. However, when planting them, care must be taken in their management because some of them are considered invasive or weeds and this causes them to invade entire surfaces without any control.

These species must have certain characteristics for use in parks and gardens, especially their appearance, based on the colour of their leaves and the colour of their flowers, as well as their growth characteristics, in order to know to what extent they can be useful for shading purposes as well as the fact that they do not cause damage to urban infrastructure.

There is a very useful concept applicable to this type of species which is that of introduced plants, i.e. they can adapt perfectly well to adverse climate and soil conditions, as well as their adaptation to drought; however, care must be taken when handling them to protect them if necessary.

I.1 General aspects of ornamental plant management.

Reforestation is vital to mitigate environmental impacts and human culture, and is necessary, as there are areas that are deeply degraded, which ultimately leads to a strong process of erosion, and greatly helps to minimise environmental degradation.

A fundamental resource to reverse this deterioration is the development of a protective cover that allows the soil fertility to be recovered and subsequently the establishment of various plants, which will help to improve the urban image and provide the environment with products derived from physiological processes such as the release of oxygen into the environment.

In green spaces, native species, with their own characteristics such as sufficient leaf area and tolerance to low amounts of water, serve to carry out programmes that ensure the success of reforestation and at the same time allow us to deepen our knowledge about them. In this way we can also consider other utilities of the species for the population.

I.2 Characteristics of the native plants of the arid and semi-arid desert of Mexico.

There is currently a system known as "Sunset" for climatic reorganisation, used as a resource for the adaptability of plant species and varieties to a given geographical area, using data on maximum and minimum temperatures, frequency and intensity of frosts, heat accumulation and evaporative demand, In this respect, the Comarca Lagunera is considered as zone 13, which generally corresponds to semi-desert regions of medium elevation, high evaporation potential, high light intensity and low accumulation of winter cold, so that the species proposed in this proposal fall within the scheme.

The predominant soils are recent alluvial soils, light profile, with textures ranging from sandy crumb to sand, heavy clays, with poor drainage, rich in potassium, magnesium, sodium and calcium, poor in nitrogen, phosphorus and minor elements that are not very soluble.

Native plants are those that have been developed in the region, and have not been imported from places with different climatic characteristics to where they are planted for their development. The Comarca Lagunera is considered an arid region at an altitude above sea level of 1139 m above sea level with defined winters with temperatures in winter that go below 0.0 °C and in the spring and summer period with temperatures that exceed 40.0 °C, with a monthly average of 20.4 °C. The annual rainfall does not exceed 264 mm per year, without having exact periods for this rainfall, which means that for the development of plants, irrigation systems must be designed for water supply, preferably with treated waste water.

I.3 Cold damage due to the effect of lower temperatures in winter.

Regarding the problems of damage by cold, I will mention that this is product of the decrease of the ambient temperature caused by the season of the year that corresponds to the winter, in what concerns to the Comarca Lagunera the winters are considered defined, this is that they occur from the month of December to the month of March, early frosts are those occurring before the onset of winter and late frosts are those occurring after the winter has elapsed, the most damaging to plant species are the late frosts because they are those in which the plants are already sprouted or are about to sprout, this is valid for deciduous plant species or species adapted to the region that can tolerate temperatures below zero degrees Celsius.

I.4 Physiological mechanisms of low-temperature tolerance in plant species

The tolerance of plants to low temperatures, both deciduous and evergreen, which are those that shed leaves or keep leaves in winter respectively, is defined by their tolerance to low temperatures (below zero degrees centigrade), This tolerance in turn is considered to be characteristic of the species and this is a product of their physiology, in terms of their cellular plasma not freezing at temperatures below zero degrees, as a result of their high osmotic potential, which in colloquial terms means the amount of total solids contained in their cellular juice. The higher the concentration of soluble solids in the cell, the more resistant it will be to cold because its freezing point is well below zero degrees Celsius, which could be from -10C to -200C, depending on the species and its adaptation to the environment. In the Comarca Lagunera recently the lowest temperature was recorded at -8.50C, which gave us an idea of which plant species could be highly tolerant, moderately tolerant or not tolerate temperatures below 0^0 C, such species from native, Table 1, introduced Table 2, of riparian origin (Rio) Table 3, fruit trees Table 4, and palm type Table 5, which are found in the Comarca Lagunera are shown in this paper, this to give us an idea especially from the point of view of urban use in parks, gardens, avenues, streets, etc..

I.5 How cold damage is caused to plant species.

Water freezes at 0^0 C, this is when it is a condition of distilled or purified water, that does not contain suspended solids, neither organic nor inorganic, and when it contains solids, the freezing point may change. In plant species the cytoplasm or cell juice regularly contains suspended solids, so that the damage in the species varies from 12^0 C to -20^0 C, depending on the adaptability, condition of their foliage or their physiology, in species originating from desert and semi-desert zones the freezing points are regularly below 00C.

Once the freezing point occurs at the cellular level due to the effect of the decrease in temperature, which can be either due to a condition of northern winds or due to the effect of a drop in atmospheric pressure, which in the first case are called white frost and in the second case black frost, the H2O molecule, i.e. water, crystallises forming a kind of "V" with the central point being oxygen and the vertices being the two hydrogen molecules, At an angle ranging from 45° to 90°, it forms a crystal with sharp edges, causing, when the process of accession takes place, the mobility of these edges causing a rupture of the cell walls considered as a PLASMOLISIS, which causes a functional disorder of the cell and consequently the death, then the tissues and finally the organs, and therefore of the plant species. This phenomenon occurs after the cold has passed and the ambient temperature begins to rise, i.e. when the cold is intense it is not harmful until the temperature rises or the cells come into contact with the sun's rays and die.

The way in which we can see probable cold damage is simply to make staggered cuts in the bark of the trees, and where the tissues are brownish-brown, the damage has already been done, and where the tissues still retain the normal green colour of the plant species, the damage has already been done.
tissue is still alive.

CHAPTER II

II. List of ornamental species, suitable for transplanting in urban areas

No	Common Name Plants Native	Scientific Name
1	Acacia salicina	***Acacia salicina***
2	Acacia agujeta	***Acacia stenophylla***
3	Ahuehuete or Sabino	***Taxodium mucronatum***
4	Silver poplar	***Populus alba***
5	Carob	***Ceratonia siliqua***
6	Anacahuita	***Cordia boissieri***
7	Casuarina	***Casuarina equisetifolia***
8	Ebony	***Diospyros texana***
9	Cypress	***Cupressus semprevivens***
10	Red Oak	***Quercus agrifolia***
11	Evergreen oak	***Quercus spp***
12	Eucalyptus	***Eucalyptus oblicua***
13	Ficus	***Ficus spp***
14	American ash	***Fraxinus americana***
15	Common ash	***Fraxinus excelsior***
16	Gravilia	***Grevillea robusta***
17	Huizache	***Vachellia farnesiana***
18	Tears of St Peter	***Tecoma stans***
19	Indian laurel	***Ficus microcarpa***
20	Lila	***Melia azedarach***
21	Mesquite	***Prosopis juliflora***
22	Chilean mesquite	***Prosopis chilensis***
23	Foreign mesquite	***Prosopis glandulosa***
24	Wicker	***Chilospsis linearis***
25	Blackberry	***Morus spp.***
26	Palo blanco	***Prosopis alba***
27	Palo verde	***Parkinsonia spp***
28	Cow's foot	***Bauhinia forficata***
29	Pinabete	***Cupressu arizonicas***
30	Pinguico	***Ehretia tinifolia***
31	Pinon pine	***Pinus cembroides***

32	Pirul	***Schinus molle***
33	Chinese Lollipop	***Schinus terebinthfolius***
34	Willow	***Salix lasiolepis***
35	Weeping willow	***Salix babyilonica***
36	Tabachm	***Delonix regia***
37	Tree-like thunder	***Ligustrum lucidum***

CHAPTER III

III. Description of Ornamental Plant Species for Urban Areas

No.1

a) **Common names**;

Acacia, and willow-leaved acacia, coba, native willow, and black acacia.

b) **Scientific name**;

Acacia salicina

1. Common name: AcaciaScientific name: *Acacia salicina*

c) **Description of the species**; It is a plant native to Australia, is commonly known as Acacia and casually is called willow leaf, this because the leaves and branches hang like a weeping willow. is a tree of considerable height, evergreen that can grow to about 15 meters high with a longevity of up to 50 years, its flowering begins in early October until January, forming pods that are those that contain the seeds, these being brightly colored black with a crimson hoop.

d) **Soil characteristics for this species**; One use for this species is that it can be planted on river banks for the purpose of stabilising the process of water erosion.

e) **Water requirements**; it can develop in regions where rainfall fluctuates between 375 and 550 mm annually, being necessary especially in its first years constant irrigation of stationary water, and when these species grow they are resistant to low moisture content in the soil, so that they can develop and maintain themselves with the aforementioned rainfall.

f) **Resistance to low and high temperatures**; tolerates temperatures up to -6 and 7 degrees centigrade (2), tolerant of high temperatures with high light intensity

g) **Recommendations for planting in urban areas**; As it is a species that once developed, its water needs are low, it is widely recommended for reforesting urban areas and for use in parks and gardens, due to the excellent shade that it provides and also because it is a plant that is visually very attractive as it resembles the characteristics of a willow tree, also in various regions of arid and semi-arid areas it is known as desert willow. Acacia is excellent for landscaping in dry areas (9).

h) References:

2. Garden at cerefree town center 2008. List of plant identification.
Wayback Macine.

9. Acacia salicina (http/www.csu.edu.su/herbarium/acacssali_sws.html)
Csu.edu.au

No.2

a) Common names; Acacia agujeta, Shoelace Acacia

b) Scientific name;

Acacia stenophylla

2. Common name: Acacia agujeta Scientific name: *Acacia stenophylla*

c) Species description; It is considered an evergreen tree native to Australia. Its plant size varies from shrubby to sprawling (4), reaching heights of up to 20 m. The bark is dark grey to blackish with smooth wax-reflecting and sometimes angular branches. The bark is dark grey to blackish with smooth wax-reflecting and sometimes angular branches. The leaflets are elongated, strap-shaped and vary from 15 to 40 cm and 2 to 10 mm in diameter. It generates raceme-like inflorescences, where the flower heads are creamy white or pale yellow. Once the flowers have been fertilised, elongated pods up to 30 cm long and about 10 to 12 mm wide are formed with a leathery, leathery texture, flowering takes place in autumn.

d) The soils to which it has adapted perfectly are those of fine texture, of alluvial origin, with clays rich in iron, it tolerates high pH, i.e. above 7.0, with characteristics of salinity, although this characteristic is not very appreciated by the species as it is not very tolerant to salts, it adapts perfectly to the soils of arid and semi-arid areas of the world.

e) Water requirements: This species, due to its leaf type constitution, requires little water, however, during the first years of growth it is necessary to water it up to twice a week and once it has adapted to the climatic conditions, watering is spaced out for a longer period of time, this is an important factor because it is regularly found in arid and semi-arid areas of the world, although it is scarce. It needs about 400 mm per year (8).

f) Resistance to low and high temperatures; This species has developed naturally in hot, arid climates, although it has been found to grow in sub-humid areas as well. It tolerates temperatures of 35 to 49 degrees Celsius in the extreme south, and low temperatures of between 4 and 7 degrees Celsius, temperatures below zero degrees Celsius can be tolerated as long as they are not for prolonged periods.

g) Recommendations for planting in urban areas; For urban areas it is recommended as an ornamental plant, but only for its appearance, as it does not regularly provide a great deal of shade due to its leaf characteristics, however, it is appreciated for the colour of its leaves and its showy inflorescences, it does not regularly raise the pavements and does not damage the drainage and conduction systems of wiring and electricity networks.

h) References:

4. Flora of Victoria 2020. Vicflora.rbg.vic.gov.au.

Pedley, L. 1992. Corrigenda-A further note on acacia aneura (Mimosideae): leguminoseae).

No.3

a) **Common names**;

Sabino, ahuehuete, Mexican cypress, Montezuma cypress, and ahuehue

b) **Scientific name**;

Taxodium mucronatum

3. Common name: SabinoScientific name: *Taxodium mucronatum*

c) Description of the species; the ahuehuete in the Nahuatl language means "an old water tree", being an arboreal species, it is native to Mexico, southern Texas and northeastern Guatemala, (3) it is considered a national tree in Mexico, for its beauty, longevity, large dimensions and traditional, as well as sacred qualities where it has been part of legends and history. They are long-lived trees, reaching hundreds of years of antiquity, they are leafy trees, with trunks of considerable diameters, between 2 and 14 metres, with some reaching heights of up to 40 metres. The leaves are arranged in a spiral, 1 to 2 cm long and 1 to 2 mm wide, and produce their seeds in pines between August and November. They usually inhabit the banks of rivers and lakes, streams and streams submerged in water. They regularly live at altitudes of 300 to 2500 metres above sea level, although they can also grow at higher altitudes (1 and 2).

d) Characteristics of the soils for this species; they regularly grow in alluvial soils, that is to say, soils of alluvial origin of clay dragging towards the streams, these are high in fertility and rich in organic matter, with considerable quantities of sandstones, as well-drained and deep soils are required, this because they reach considerable heights.

e) Water needs; they regularly grow in places where water conditions are abundant, it is rare to find this type of tree outside their habitat which is between rivers and lakes, so their water needs are high. They can be found in parks and gardens but require a lot of water to thrive and grow.

f) Resistance to low and high temperatures; Temperature conditions are variable, tolerating temperatures up to 40 degrees Celsius and low temperatures down to -10 degrees Celsius, as they always contain constant high water concentrations.

g) Recommendations for planting in urban areas; Only if there is enough space for their prolonged growth, and sufficient water to keep them always with high levels of hydration, they are trees with abundant shade and an excellent visual presentation. They are not recommended for pavements because of their foliage volume and heights.

h) References:

1. International Union for Conservation of Nature 2022. Taxodium mucronatum.

2. Sternberg, G; James Wesley Wilson 2004. Native trees for North American Landscapes.p.476.

3. Frailish, J.S. and Scott, B. Franklin. 2022 Taxonomy and Ecology of Wood plants in North American Forests. P. 123.

No.4

a) Common names;

Silver poplar, Afghan poplar, white poplar, white poplar, common poplar, white poplar

b) Scientific name;

Popuhlus alba (1)

4. Common name: Silver Poplar Scientific name: *Popuhlus alba*

c) Description of the species; Native to the arctic Palearctic its natural distribution is in the Iberian Peninsula and Morocco and it is regularly found in most of Europe, it can also grow in temperate zones of South America and Mexico. It is considered a deciduous tree with long and rounded growth, fast growing up to 3 m in height and generating trunk diameters of up to 1 m. It has a strong root system. The bark is smooth greyish-white, with blackish scars on old branches. The leaves are simple deciduous, alternate oval and palmate, with toothed edges, covered on the underside with a dense layer of whitish felted hairs, the young ones white and hairy and the adults with a dark green upper side and white underside, with a very polymorphous blade. The seeds are rounded or oval, slightly lobed. The catkins are hanging, the male catkins are 3 to 6 cm long and the female catkins are longer and thinner. The male flowers are reddish in colour and the female flowers are greenish-yellow. The fruits are ovoid bivalve capsules and lampinas.

d) Soil characteristics for this species; they grow perfectly in loamy textured, rich in fertility and humid soils, in the vicinity of rivers, however, they are able to grow in sandy soils and soils that support waterlogging, due to sea water in their root system, they can also develop in poor and calcareous soils.

e) I lid rich needs; they require significant amounts of water, even withstand salty water, i.e. being close to the sea, these species are used as defence screens near the sea.

f) Resistance to low and high temperatures; withstands cold well, as long as the temperature is not low and for a long time, also supports excessive heat, just do not miss the water.

g) Recommendations for planting in urban areas; Its root growth is excessively large and strong, therefore it should not be planted on pavements or near buildings and cable installations. They are cultivated as ornamental trees, but they need large gardens in avenues, because of their pleasant appearance due to their colour changes, from green to white simulating a silver colour and the good shade they offer,

h) References:

1. Charles Linnaeus. 1753. Description of the species. Species Plantarum. 2:1034.

No.5

a) Common names; carob, white carob, carob cacha, feral carob, etc. ....

b) Scientific name;

Ceratonia siliqua

5. Common name: Carob tree Scientific name: *Ceratonia siliqua*

c) Description of the species; The names are born from the figure of water this The fruit has a horned epithelium, which means horn in Greek. It is a tree up to 1 metre high, varying in height from 5 to 6 metres, it is dioecious with evergreen foliage. It has dark green paripinnate leaves 10 to 20 centimetres long. The flowers are small and appressed, of radiate symmetry, yellowish green colour in calcogenous and erect clusters that come out of the oldest leaves of the tree. It is quite hardy and resistant to drought, slow to develop and reaches adult maturity at seven to ten years, being fully mature at fifteen to twenty years. Its fruit is a pod (7), coriaceous of dark brown colour of about 3 cm in length, these contain a gummy pulp of sweet and pleasant taste and are used as fodder. These fruits can be harvested when the mature trees produce up to 90 or 200 kg, and are harvested in August (13).

d) Soil characteristics for this species; it requires rustic soils, i.e. low fertility and shallow soils. Due to the nature of its growth.

e) Water needs; tolerate prolonged drought, i.e. they do not need much water for their growth and development.

f) Resistance to low and high temperatures; Although it is a perennial tree, care must be taken to protect them from the cold at an early age and when they are in full growth they can tolerate temperatures below zero degrees centigrade, although not for prolonged periods.

g) Recommendations for planting in urban areas; It is recommended for regions with low relative humidity and tolerates long periods of drought, austere soils. This is without much fertility and low in organic matter, they are showy trees, although of medium height for its intense green colour of the leaves, gives good shades. They are also recommended for regions with low temperatures, because of their tolerance when they are already developed. They also thrive in warm regions with good growth and good shade conditions.

h) References:

7.Gaffiot F., 1934. Dicctionnaire latm-Francais Hacheette Paris. P. 1442.

Ceratonia siliqua . 2009. Plantasutiles.
(http://www.linneo.net/plut/index2.htm)

No.6

a) Common names;

Anacahuita, Anacahuite, Cueramo, Rasca Viejo, trompillo and wild olive tree

b) Scientific name;

Cordia boissieri

6. Common name: AnacahuitaScientific name: *Cordia boissieri*

c) Description of the species; It is a small tree or shrub that can reach a maximum height of up to 6 metres, it has perennial leaves, with thin bark, its leaves are oval reaching up to 12 centimetres in length. Its flowering begins at the end of spring and until the beginning of summer, even its flowering in some regions can take place all year long, being these flowers of white colour, with its interior of yellow colour, they are big giving a pleasant appearance at the beginning of spring. The fruit is round and greenish-yellow in colour, similar to the olive tree, hence its name of wild olive tree, about 2.5 cm in diameter and contains four seeds. The fruit is slightly toxic and can cause dizziness. It is widely distributed in regions of Mexico such as La Sierra Madre Oriental, Nuevo León, Tamaulipas, San Luis Potos^ and Hidalgo in Mexico (1).

d) Soil characteristics for this species; It can develop

perfectly on light calcareous rustic calcareous soils, i.e. low fertility,

e) Water requirements: Tolerates prolonged drought, this is one of its main characteristics, however, it is important to take care in its early years to supply it with the necessary moisture,

f) Resistance to low and high temperatures; tolerates low temperatures, but not for prolonged periods below zero degrees Celsius, and tolerance to temperatures above 40 degrees Celsius can also be tolerated, but it is important to provide water when this is a continuously recurring climatic condition.

g) Recommendations for planting in urban areas: Due to its tolerance to low and high temperatures, it is recommended for backyard and reforestation in urban areas, as they are also pleasant to look at when flowering and are not very tall. For the case of the radical system we will not have any problem with raising pavements or affecting the drainage system and wiring.

h) References:

1. CONAVIO. 2009. Catalogo taxonomico de especies de Mexico. Mexico City. In Capital Nat. Mexico.

No.7

a) Names

common; Casuarina (1), Australian Pine, Pansy Pine, Sadness Tree, Horsetail Casuarina.

b) Scientific name;

Casuarina equisetifolia

7. Common name: Casuarina Scientific name: *Casuarina equisetifolia*

c) Description of the species; Its height reaches up to 25 or 30 metres, with The bark is divided into longitudinal bands, and has a lot of tannins. Its foliage makes it look like a cornifer, but it is not, as it has thin leaves, similar to those of the pine aticules, but the difference is that they are septate, its stems are green, similar to those of the species of equisetum, measuring 10 to 20 centimetres in length, The flowers are unisexual, not very showy and small in size, this species is monoecious, as it has male and female flowers in different parts of the plant, and when they are fertilised they form a false pineapple, globose, which contains the fruit, the size of which varies from 5 to 8 mm in diameter. It multiplies by seed very well, being the seeds viable for up to 2 years, being a tree that reaches great height.

d) Soil characteristics for this species; It grows on sandy xeric shores, salty, calcareous soils, as well as in the mountains on heavy areas, on rain soils with volcanic characteristics, as well as on marginal soils (3). The plants are very well anchored to the soil and have allelopathic properties.

e) Water requirements: Being an austere species tolerating poor to alluvial soils, it can tolerate droughts, although these should not be so prolonged.

f) Resistance to low and high temperatures; it tolerates low temperatures but not for a long time, especially when the trees have been in the same place for more than 5 years. It also tolerates high temperatures, although for a short time, due to its characteristic of being similar to cork trees, which develop at high altitudes.

g) Recommended for planting in urban areas; it is very useful for reforesting rural and urban areas in the tropics, subtropics and temperate regions. It fixes atmospheric nitrogen like a legume and is a fast-growing tree. It has mycorrhizae or fungi association with the r;u'z, having a fast growth due to the fixation of atmospheric nitrogen. It is important to mention that for its reforestation it is necessary to be very careful since the branches of this species are very fragile and can cause damages to the pedestrians, especially in seasons of intense air or tolvaneras.

h) References:

1 Valdes, M., A.C. Rodrigo; M. A. Leyva; and A.D. Camacho. Rodrigo; M. A. Leyva; and A.D. Camacho. 2004. Growth promotion in ***Casuarina equisetifolia*** nursery by symbiont microorganisms. Chapingo, Mexico.

3 . C. Y. Wu, P.H. Raven and D. Y. Hong 1999. Flora of China Editorial Committee. China Science Press and Missouri Botanical Garden Press. 4:1-453.

No.8

a) Common names;
Ebony, chapote, chapote manzano, or chapote prieto.
b) Scientific name; ***Diospiros texana.***
8. Common name: Ebony Scientific name: *Diospiros texana*

c) Species description; native to central and west Texas USA, as well as western Chihuahua, Coahuila and Nuevo Leon (3). It is a large tree or shrub (2), with a lifespan of 30 to 50 years, it can reach up to 12 m in height, its bark is light reddish grey, with sclerophyllous leaves, they are oval, dark green, 2 to 3 cm wide, the lower surfaces are covered with fine hairs, and the upper surfaces are bright green, it has branches with thorns. The trees are deciduous, becoming evergreen species. It is a dioecious plant, producing white flowers from March to April. It is a dioecious plant, producing its flowers from March to April, they are white in colour and are skinned from female catkins and bear fruit in berries which are very hard.
d) Soil characteristics for this species; it grows in riverine meadow margins and rocky slopes, preferring well-drained, alkaline soils.
e) Water requirements: Tolerates long periods of drought, but in its early stages it is important to keep the soil moist while it is developing, and once it has reached maturity, watering can be delayed for up to a month, so that it can develop adequately.
f) Resistance to low and high temperatures; tolerates temperatures below zero degrees Celsius, but not for prolonged periods, and temperatures above 40 degrees Celsius are tolerable for this species as long as they have an adequate

moisture content.

g) Recommendations for plantations in urban areas; It is a striking species for its bright green colours of its leaves, although its branches have thorns especially in its juvenile period, they give a great shade, they are not very clean species because the birds prefer them to make their nests always generate this dirt, however, by giving a good surplus if they are required to reforest both urban and rural areas.

h) References:

2. University of Texas at El Paso.2006. Texas persimmon. Chihuahua Desert Plants.

3. Watson, George, 1938. Nahuatl Words in American English. American Speech 13 (2): 113-114 pp.

No.9

a) Common names;

Cypress, common or Mediterranean cypress

b) Scientific name;

Cupressus sempervirens

9. Common name: CypressScientific name : *Cupressus sempervirens*

c) Description of the species; It is an evergreen tree species of the

It is also known in the United States as Italian cypress, because it is very common in Italy, although it is not originally from that region. It grows to a height of 25 to 30 metres, but can reach up to 42 metres. Its leaves are presented in small branches in the form of scales between 2 to 5 mm in length, forming a dense foliage of dark green colour. The branches are thin, cylindrical and dark green in colour. The fruits are small greenish-grey pines or cones, 2 to 3 cm in diameter, which take on a woody appearance when ripe. The cones develop in spring and ripen in the autumn of the year following pollination, about 20 months later. The trunk is straight and reaches up to 1 metre in diameter and with some exceptions, trees up to 3 metres in diameter have been found. The bark is thin and smooth, greyish grey in colour in young trees and with age it changes to dark brown with longitudinal cracks. Its rafces are horizontal, superficial and elongated, which are the secondary ones. Its structure is horizontal with extended ramifications giving a cedar or pine appearance, pyramidal, with a columnar shape. It is also considered a medicinal plant and is used in infusions as a treatment for varicose veins, varicose ulcers, haemorrhoids and prostate problems (4).

d) Soil characteristics for this species; tolerates both acidic and alkaline soils, withstands sandy, compacted soils, rejects wet or sandy soils.

e) Water requirements: As it develops in Mediterranean areas, it requires constant humidity, although it is not very tolerant to drought, with rainfall in summer for its normal development.

f) Resistant to low and high temperatures; thrives in regions with hot summers and harvests in wet winters. It can thrive in cold areas with more humid summers, with temperate subtropical high altitude climates, with summer rains as in Mexico. Due to its funerary symbology, cemeteries usually have the best specimens. It tolerates very well the heat and the shade, where the tree needs sun. It tolerates very well cold down to minus 15 degrees centigrade, with some adult trees tolerating lower temperatures.

g) Recommendations for planting in urban areas; It is used as an ornamental tree because of its pyramidal structure, and is also used as a windbreak because of its firm, upright structure. Due to its size it is used for decoration in parks and gardens.

h) References:

4. Idarraga-Piedrahfta, A., R. D. C. Ortiz, R. Callejas Posada & M. Merello (eds.) 2011. Fl. Antioquia: Cat. 2: 9-939. University of Antioquia, Medellm

No.10

a) **Common names**; Red oak, California oak (1) or coastal oak

b) **Scientific name**;

Quercus agrifolia

10. Common name: Red oakScientific name: *Quercus agrifolia*

c) **Description of the species**; It is classified in the red oak section. This species is green in the spring seasons until late summer, then due to its deciduous condition it turns red and then brown until it becomes completely bare in late autumn and early winter, when temperatures are low and photoperiods are short. At maturity it reaches heights of 10 to 25 metres, some specimens can reach an age of 250 years, with trunks of up to three or four metres. Its leaves are green, curved, similar to the Canadian maple, 2 to 7 cm long and 1 to 4 cm wide. The flowers are produced in early to mid spring, the fruit is a thin brown acorn, these mature between 7 to 8 months after pollination, we can find them in high latitudes to more than 2000 meters above sea level, this limits it to develop at lower altitudes with the restrictions of growing more slowly.

d) **Soil characteristics for this species**; it is a forest-dwelling species, it develops perfectly in well-drained soils on hills and coastal tables, near streams, and it can also develop perfectly in evergreen forests.

e) **Water needs**; due to its forest condition, its water requirement is low, especially when the tree is developed, the first years of growth and development it does require constant water.

f) **Resistance to low and high temperatures**; grows in winter climates, not too cold, and not too hot summers, tolerating a little humid sea air.

g) Recommendations for plantations in urban areas; it is widely recommended for reforestation in both urban and rural areas, being attractive due to the type of leaves and its colour variation that goes from reddish green to brown as the months go by until winter, besides the fact that it provides perfect shade and does not damage the pavements or the root system when placed in flower boxes or backyards.

h) References:

1. Abrams, L. 1923. Ferns to Birthworts. 1: 1-557. In L. Abrams (ed.) Ill. Fl. Pacific States. Stanford University Press, Stanford

No.11

a) Names

common; Evergreen oak

Green or southern holm oak

b) Scientific name;

Quercus virginiana

11. C. name: Evergreen oak Scientific name: *Quercus virginiana*

c) Description of the species; It is an evergreen tree, that is to say that it always keeps its foliage, even if its leaves lose all their leaves and fall off, but the new leaves are already replacing the fallen ones. It is native to southern USA and northern Mexico. It reaches an average height of between 15 and 20 metres, the trunk of an adult can reach 1.2 to 2 metres in diameter. It has a rough, cracked bark (1), with deep roots, as it is propagated by seed. The young shoots have a light whitish coat, the adult branches tend to spread out horizontally, having dense foliage, forming a round or broad crown. The leaves are broad cottony and broad thick with rounded apices. It has petioles of 5mm in length with dark green glossy, shiny dark-green, curved leaf margins. It can shed its leaves in prolonged drought or long periods of cold. The male flowers appear in hanging clusters of 7 to 10 cm long, the female flowers are solitary or in groups of 2 or 3 (2). The fruits are in the form of acorns which appear in September, falling to the ground in November and December. They are oval, up to 2.5 cm long, covered with a cupule of hairy scales, arranged in groups of 3 to 5.

d) Soil characteristics for this species; it grows in either acidic or alkaline, flat, steeply sloping soils and is salt tolerant, so it can thrive in coastal forest climates, but does not tolerate sea winds.

e) Because of their forestry characteristics they can develop in places where water is scarce and are not very tolerant to drought in the subsequent years of their growth and development, however, as they are young, they are demanding of constant humidity.

f) Resistance to low and high temperatures; tolerates low temperatures as it is a forest tree, and maintains its foliage in winter even when temperatures are low, on the other hand, it tolerates high temperatures as long as it is well hydrated.

g) Recommendations for plantations in urban areas; they are recommended for backyard urban and rural areas, as they are eye-catching due to the green colour of their leaves, although they are slow growing, once they grow they generate their radical system, but they grow in depth, this makes them feasible for use in streets, as they are not dangerous for raising pavements.

h) References:

1. QUERCUS VIRGINIANA Mill. 2018.
2. VIRGINIA OAK: *Quercus virginiana*". 2018.

No.12

a) Common name;

Eucalyptus and eucalyptus trees

b) Name scientific;

Eucalyptus obliqua

12. Common name: Eucalyptus Scientific name: *Eucalyptus obliqua*

c) Species description; It belongs to the myrtaceae family,

These species are considered medicinal plants, and perennial, with a straight habit, these trees can reach up to 60 m in height, and it is mentioned in the literature that there have been specimens up to 150 m in height (7). The outer bark is usually brown, light brown with a skin-like appearance, peeling off in strips leaving grey or brownish patches on the smoother lower bark (8). The leaves are sessile, oval, greyish and falciform, elongated, bright bluish-green when adult, containing an essential oil with a characteristic balsamic odour, which is a powerful natural disinfectant. The wood can be used for fence posts, as well as for the production of paper and biofuels.

d) Characteristics of the soil for this species; Eucalyptus produces acidification of soils and decreases clay content (23).

e) Water requirements: Because they are very large species with large root systems, abundant water is recommended, especially in their early stages, and because they have a deep root system, when they are large they explore a large area of soil, so water is constantly available.

f) Resistance to low and high temperatures; these trees in their adult stage do not tolerate frost, or partially tolerate temperatures below -5 degrees Celsius, but not for long periods, although there are some species such as the snow eucalyptus which is able to withstand temperatures as low as -20 degrees Celsius, although there are also some species which can tolerate cold. These species can thrive in tropical and temperate climates, tolerating high temperatures, above 35 degrees Celsius, but when these limits are exceeded they regularly begin to dehydrate from top to bottom.

g) Recommendations for planting in urban areas; Basically, their use has intensified as a windbreak barrier, although they are usually fragile and too tall to be widely used in parks and gardens, especially because of the risk to passers-by who can cause accidents when their branches break off.

h) References:

7. Eucaypt_redemption.2008 Eastern Native Tree Society.

8. Luzar J. (2007). The Political Ecology of a "Forest Transition": Eucalyptus forestry in Southern Peruvian Andes. *Ethnobotany Research & Applications*.

Korchagin, Jackson; Bortoluzzi, 2019. Edson Campanhola; Moterle, Diovane Freire; Petry, Claudia; Caner, Laurent (2019-04). *CATENA* **175**: 132-143.

No.13

a) Common names;
Ficus, Yucatecan tree
b) Scientific name;
Ficus spp
13. Common name: FicusScientific name: *Ficus spp.*
c) Description of the species; this Genus contains about 900
species of trees, shrubs and climbers of the family ***Moraceae,*** native to intertropical zones with some of them distributed in temperate zones. Most of them are perennial. One of the main characteristics of the species of this genus is the milky secretion called latex, which is secreted when any part of the plant is cut or wounded. Another characteristic is that the terminal buds of the leaves are enclosed within a pair of stipules which are initially welded and then deciduous. Trees of this genus feed numerous species of animals and are considered keystone species in the tropical ecosystem. The false fruits are specially adapted structures called sycones, bulbous in shape with a small opening. These structures are pollinated by small fig wasps, which penetrate through the opening to fertilise them, giving tiny fruits with a hard core known as an achene. The common fig belongs to this genus, ***Ficus carica***, giving up to three crops per year of fruit palatable to livestock. The Ficus also reproduces easily by cuttings. Wasps pollinate most of these species.
d) Soil characteristics for this species; n because of its adaptability it tolerates various types of soil, however, in calcareous soils it tends to present minor

element deficiencies causing severe yellowing.

e) Water requirements; when the trees are small in size they do require water constantly, but as they develop the larger trees will tolerate some water starvation because they have a very abundant root system.

f) Resistance to low and high temperatures; being from tropical and subtropical areas, it tolerates certain temperatures above 35 degrees centigrade, however, as it develops in temperate areas it does not tolerate temperatures below zero degrees, especially for prolonged periods, causing it to die in this area and even the death of the root system. In temperate regions it is no longer growing widely because of this characteristic.

g) Recommendations for planting in urban areas; ***Ficus macrocarpa*** is used as a shade tree in the streets of many localities, in warm and subtropical climates, however in temperate zones these species are considered as an introduction, having a lot of roots in urban areas with the only disadvantage that when they grow too much they raise pavements, destroy drains and cause problems in the electric cables due to their excessive dimensions..., although they are still of good ornamental aspect due to their waxed leaves and very good shade in these parts.

No.14

a) Common names;

American ash, white ash, carolina ash, white ash, Carolina ash

b) Scientific name;

Fraxinus americana

14. Common name: American Ash Scientific name: *Fraxinus americana*

c) Description of the species; it is one of the best known of the genus ***Fraxinus***, and that of ***Fraxinus americana*** was described by Charles Linnaeus (5). It is up to 35 metres tall, being native to eastern North America. The wood is a strong, light, grainy wood. The leaves are 2 to 3 cm wide, compound, pinnate, 5 to 9 leaflets, 6 to 13 cm long, turning red or purple in late autumn. This tree is deciduous, flowering occurs in spring, the fruit is a samara, 3 to 5 cm long, with a pale brown wing, and can be blown by the wind for long distances. Longevity fluctuates around 100 years, and the wood is used for baseball bats and hand tools. The raffes are astringent, sudorific tonic, diuretic.

d) Soil characteristics for this species; they develop in any type of soil, from alluvial to calcareous, having problems in the latter as they do not develop their leaves well due to nutrient deficiencies.

e) Water needs; they tolerate certain droughts, but it is important to consider that in the early stages it is important that they do not lack humidity; these species can withstand lack of humidity, although not for long.

f) Resistance to low and high temperatures; it tolerates certain temperatures down to zero degrees, but when these are prolonged they usually cause damage to the aerial part, even reaching the skin, it supports high temperatures, above 35 degrees centigrade, although the maintenance under these conditions should be to water it constantly so that it does not dehydrate.

g) Recommendations for planting in urban areas; Due to their broad leafy leaves and good shade conditions they are widely used in green areas, especially in parks and gardens. However, it must be considered that low temperatures can cause their death, especially in regions where temperatures fall below zero degrees and for prolonged periods of time.

h) References:

5. *Fraxinus _americana*.2014. Date consulted. *Tropicos.org. Missouri Botanical Garden.*

No.15

a) Common name;

Common ash, Northern ash

b) Scientific name;

Fraxinus excelsior

15. Common name: Ash tree Scientific name: *Fraxinus excelsior*

c) Description of the species; it belongs to the oleaceae family, it is a

Tree native to most of Europe, mainly Spain, the southern part in the north of Greece. Deciduous species reaching heights of up to 40 metres, with greyish-grey bark, smooth and with lenticels on branches and young specimens, and cracked on adults. Light to dark green imparipinnate leaves, with 9 to 13 lanceolate, toothed-edged leaflets.

The leaves of this ash tree open late in spring and are the first to fall in autumn. The parnculate inflorescences, which appear before the leaves, appear in the axils of the branches of the preceding year. They are unisexual flowers, can be female, merculine or hermaphrodite, lack perianth, have two stamens with ovoid anthers. The fruit is a winged, lanceolate pod about 28-48 cm long by 5-10 mm wide, the seed is fusiform, brownish in colour (4). The wood is used in cabinetmaking, it has a good texture and straight grain.

d) Soil characteristics; for this species; This species prefers fertile soils, with good water retention, but well drained, without waterlogging.

e) Hydrological needs; requires constant humidity, but not in excess in well-drained soils, periods of drought do affect the good development of this species (2).

f) Resistance to low and high temperatures; it is a tree that tolerates cold, i.e. temperatures below zero degrees, without damaging its root system, inhabiting from 400 to 1800 metres above sea level. Although when young it can tolerate some shade, it needs to be exposed to the sun,

g) Recommendations for plantings in urban areas; This species

it is cultivated as an ornamental, due to its high flexibility and resistance to splitting, in addition to giving good shade it has a pleasant appearance, although the leaf is smaller than the American brake, it is recommended in parks and gardens for this characteristic as well as not drastically damaging the pavements as it has a deep, pivoting root system.

h) References:

2. Autoecolog^a del fresno comun, Date of consultation 2017.

4. Floraiberica.es. Date of consultation 2017

No.16

a) Common names;

Gravilia, silky oak, silver oak, fire tree, and gold pine

b) Scientific name;

Grevillea robusta

16. Common name: Gravilia, Scientific name: *Grevillea robusta*

c) Species description; Fast-growing evergreen tree, 15 to 35 m high with dark green, delicately toothed, bipinnate leaves, reminiscent of fern fronds, 15 to 30 cm long, with a slightly greyish-white underside. The flowers are golden-orange, 8 to 15 cm long in spring on 2 to 3 cm stems. The seeds fructify in dark brown velvety dehiscent follicles, 2 cm long with one or two flat winged seeds. It is commonly used for musical instruments such as guitars, including fine furniture. The flowers and fruits contain the toxic hydrogen cyanide (5), and care should be taken because contact can cause contact dermatitis (6).

d) Characteristics of the soils for this species; It tolerates calcareous soils, these with high content of carbonates, it needs deep and very fertile soils this you make that when you have its leaves they are noticed but shining.

e) Water requirements: This species occasionally needs water, being resistant to drought.

f) Resistance to low and high temperatures; they grow in temperate zones, however, they need protection from night frosts, when they are at a young age, tolerating temperatures down to -8 degrees Celsius.

g) Recommendations for planting in urban areas; due to its large size and having some resistance to rot it is recommended for urban parks and gardens, having excellent shade and a pleasant appearance, especially in the flowering season, which is in early spring.

h) References: 5. Verdcourt, B.; Everist, S. L. (1976). "Poisonous Plants of Australia. *Kew Bulletin* **31** (1): 191.

6. Menz, Jennifer; Rossi, Ric; Taylor, Wal C.; Wall, Leon (1986-08). "Contact dermatitis from Grevillea 'Robyn Gordon'". *Contact Dermatitis* **15** (3): 126-131.

No.17

a) Names
common; Huizache,
aromo and guisache,
b) Scientific name;
Vachellia farnesiana
17. Common name: Huizache, Scientific name: *Vachelliafarnesiana*

c) Species description; Native to tropical America, and cultivated all over the world (2), it is considered a thorny, small, evergreen sub deciduous shrub, 1 to 2 metres high, shrub-like, 3 to 10 metres in form Arborea, rounded crown, bipinnate leaves, alternate, with a pair of straight stipulate thorns, not thickened basally, 1 to 2 cm long, total size 2-8 cm, The trunk is short and slender, well defined, abundantly branched from the base, with ascending and sometimes horizontal branches, with smooth outer bark when young and fissured when old, lead-grey to dark brownish-grey, with abundant lenticels. Internally it is yellowish-cream coloured, fibrous, with a marked garlic smell. The flowers are yellow glomeruli, originating from the axils of the spiny stipules, solitary or in groups of two to three. The fruit is a legume, reddish-brown, hard, 2 to 3 cm long, ending in a sharp point. With dehiscent leaflets that remain on the tree after ripening. The seeds are chestnut-coloured, arranged in two rows, immersed in a spongy, whitish pulp, with a waterproof head (4).

d) Soil characteristics for this species; it develops perfectly in any type of soil, thriving well in dry, saline and even sodic soils,

e) This species is resistant to drought, although it is not resistant to drought. development depends basically on good moisture content,

f) Resistance to low and high temperatures; it can tolerate low temperatures down to below zero degrees, on the other hand, it does require abundant heat for its optimum development.

g) Recommendations for plantations in urban areas; it is used in
In gardening as an ornamental, and in beekeeping for its abundant flowers, also in live fences for its resistance and the amount of thorns that it develops up to 15 cm long. In urban areas it is recommended for its surplus, they develop deep branches but its thorns are a disadvantage.

h) References:

2. *Acacia farnesiana* (L.) Willd. at Purdue University, CNCPP
4. *Acacia farnesiana* in Flora Iberica, RJB/CSIC, Madrid *pro parte*
Jump to[a b]

No.18

a) **Common names**; Lagrima de San Pedro, tronador or tronadora.

b) **Scientific name**;

Tecoma stans

18. Common name: St. Peter's Tear Scientific name: *Tecoma stans*

c) **Description of the species**; Found in a wide variety of environments on almost every continent, (1). It is a small, hermaphrodite perennial shrub or tree, with hard wood and compound, opposite, serrated-edged leaves. The fruit is elongated, 7 to 21 cm, brownish-green in colour. The main characteristic is the flower with a bright yellow, 3 to 5 cm. tubular, bell-shaped corolla (2). The branches and also the trunk are split lengthwise.

This makes the tree unsafe, especially for children's playgrounds and parking areas. It is common to see dry branches hanging from the tree, as well as branches that break off and continue to live.

d) **Soil characteristics for this species**; Because of its invasive character it grows in any type of soil from sandy to rocky.

e) **Hydrological requirements**; tolerates prolonged drought and thrives in any environment,

f) **Resistant to low and high temperatures**; grows from temperate forests to tropical deciduous and evergreen forests. To xerophytic scrub and intertropical coastline (2).

g) **Recommendations for planting in urban areas**; It is cultivated as an ornamental plant, for its showy yellow flowers, it has a wide variety of uses, and more than 50 chemical components. It is widely developed for its beautiful blooms, to adorn streets and gardens. It has an invasive potential, and occasionally becomes a weed, quickly colonising rocky, sandy and cleared fields.

h) References: 1. Tecoma stans". *Enciclovida.mx*. Conabio. Accessed 17 February 2020.

2. Jump to:[a b] CONABIO. "Tecoma stans". Accessed 8 January 2017.

No.19

a) Common name;

Indian laurel, Yucatecan

b) Scientific name;

Ficus macrocarpa

19. Common name: Indian Laurel Scientific name: *Ficus microcarpa*

c) Description of species; native to south and south-east Asia,

It is considered an invasive species in these regions (2). In Mexico it is known as the yucatecan tree. It is

a perennial tree of good size and rapid growth, reaching up to 15 metres in height, very branched and with a voluminous crown. The leaf is dark in colour with a coriaceous appearance, it is very shiny, it can measure from 4 to 13 cm in length, it has white flowers that emerge from the axils of the leaves, they are dioecious and produce small fruits of 1 cm in diameter, called sycones of green colour turning yellowish to reddish when ripe (3). A wasp known as the fig wasp is involved in pollination.

d) Soil characteristics for this species; it adapts to all types of soils, especially deep and fertile soils in temperate or calcareous regions. In the first years it suffers from certain deficiencies, especially of minor elements, but as it grows these deficiencies are no longer a problem.

e) They require constant water, especially in their early stages of development, and as they grow, due to their large area of root exploration, they tolerate certain degrees of drought.

f) Resistance to low and high temperatures; being a species of tropical and subtropical climates, it is very sensitive to low temperatures, especially when these are very low and for prolonged periods, tending to die even at the root level. High temperatures are very favourable, as they help it to grow rapidly.

g) Recommendations for planting in urban areas; used as a shade tree. In streets of many localities with warm and subtropical climates, in temperate climates it tends to grow to great heights with exuberant shadows, however it does not tolerate temperatures below zero degrees and tends to die to the root, on pavements when they grow too much they tend to lift it up, causing damage to drainage systems. Because of their large growth they are not advisable in the passage of light cables, unless a good programme of constant pruning is in place. It is also cultivated as an ornamental plant for interiors with its respective management, including as a bonsai tree.

h) References:

2. Plants of Hawaii - hear.org

3.$ FLEPPC (http://www.fleppc.org/index.cfm Florida Exotic Pest Plan Council Archived 24 May 2011 at the Wayback Machine.)

No.20

a) Names

common; Lila,
cinnamon, agriaz,
piocha, canelo, pnrniso sombrilla, or p;ir;n'so tree.

b) Scientific name;

Melia azedarach

20. Common name: Lilac Scientific name: *Melia azedarach*

c) Description of the species; It is a deciduous tree of medium size, growing from 8 to 15 metres high, with a straight and short trunk. The crown reaches 4 to 8 metres in diameter, in the shape of an umbrella, the leaves are opposite, compound, with long petioles, imparipinnate, from 15 to 45 cm in length, The leaflets are oval, acuminate, 2 to 5 cm long, dark green on the upper side and light green on the underside, with a tapering margin, the flowers are pentamerous, purple or lilac in colour, arranged in terminal panicles up to 20 cm long and are very fragrant. The fruit is a drupe, 1 cm in diameter, globose in shape, green and pale yellow when ripe, with a thick endocarp, with one seed per locule, 1 mm long by 0.3 mm wide. The fruits have narcotic properties, causing toxicity of the fruits that can affect humans (5), (6).

d) Soil characteristics for this species; It adapts perfectly to acid or alkaline soils, and tolerates a certain degree of salinity in the soil.

e) Water requirements; tolerates drought, although not very prolonged, however, it is important to maintain a good moisture content, especially during the growing season.

f) Resistant to low and high temperatures; grows favourably in warm and temperate areas all over the world, tolerates light frost, provided there is a warm summer.

g) It is cultivated as an ornamental in South Africa and America, where it naturalised quickly, becoming an invasive species, which displaced other autochthonous species. It is used for decoration and shade, mainly because of its broad leaves and leafy crown, which is the reason for its common name. It is used in gardening as a shade tree, highlighting its abundant aroma and flowering.

h) References:

5. Little, Elbert L. (1994). Knopf, ed. *The Audubon Society Field Guide to North American Trees: Western Region*. Chanticleer Press. p. 517.

Phua, Dong Haur; Tsai, Wei-Jen; Ger, Jiin; Deng, Jou-Fang; Yang, Chen-Chang (1 January 2008). "Human Melia azedarach poisoning". *Clinical Toxicology* **46** (10): 1067-1070.

No.21

a) Common name;

Common mesquite,

b) Scientific name;

Prosopis juliflora

21. Common name: Mesquite Scientific name: *Prosopisjuliflora*

c) Description of the species; it is a species of leguminous plants, of zones of dry tropical forest, being this a thorny tree, of deciduous behaviour growing until reaching in conditions of good soil and suitable climate up to 20 meters of height, of irregular crown and of sparse foliage, extended, of compound leaves, alternate, pinnate, bipinnate, of 10 to 20 cm of length, with petioles widened from the base of 2 8 cm composed of 12 to 16 pairs of folioles by leaf, of 20 to 22 mm of length. Irregular stem bifurcate from the base, with thorns on the young branches (3). Inflorescences of yellow colour, in spiky cylindrical clusters, 5 to 8 cm long, Contains slightly leafy fruits, indehiscent, 11 to 12 cm long, by 0.8 to 12 mm thick, purplish yellow, with longitudinal reddish striations, its size ranges from 6 to 9 to 6 mm long by 4 to 6 mm wide. Contains flattened seeds surrounded by a sweet membrane. It is a very efficient, fast-developing, phreatic-rooting plant. It is hermaphroditic, being these plants self-incompatible, so their interbreeding is necessary. It flowers regularly from December to February in subtropical areas, while in temperate areas it flowers in spring, being entomophilous pollinated, and can be propagated by seed, rootstock, air layering, cuttings and grafting. Its wood in arid and marginalised regions in recent years has been widely used to make charcoal, which is a source of subsistence for the population living in these conditions.

d) Soil characteristics for this species; it develops perfectly in deep soils and tolerates significant amounts of sand, as its root system is very deep, fast developing, able to take advantage of the water available in the subsoil. It grows in various types of soil, even poor soils with a pH of 6.5 to 8.3. It requires little rainfall. This tree tends to improve soil fertility, controls erosion and nitrogen fixation, as it is a legume.

e) Water requirements; In arid soils it develops its inizia at great depths of up to 20 metres, with rainfall from 150 to 250 mm per year. It grows from 0.0 to 1500 metres above sea level.

f) Resistance to low and high temperatures; withstands high temperatures between 23 and 29 degrees Celsius. Even with intense sunstroke (4). It is resistant to frost, although it dies at temperatures of minus 4 degrees Celsius.

g) Recommended for planting in urban areas; regularly used for shade and its hardwood is widely used in carpentry. It provides shade and food for wildlife and domestic animals. It is not recommended for parks and gardens because of its ashy appearance and large number of thorns, although it provides good shade. Its pods are edible for animals and humans and are an important source of food in marginalised regions.

h) References: 3. "Prosopis juliflora". *conabio*. Accessed 12-11-2022.

4. Jump to: "El cuji yaque". *PDVSA Boletm ecologico.*

No.22

a) Common names;
Chilean mesquite , or
tamarugo

b) Scientific name;
Prosopis chilensis

22. Common name: Chilean mesquite Scientific name: *Prosopis chilensis*

c) Description of the species; ***Prosopis chilensises*** (1) a South American Arborea species, leguminous (2) of the family of those that inhabit the North central zone of Chile, being located between 500 to 2500 meters above sea level. It is a medium low tree, no more than 3 to 12 metres high and 6 metres in diameter when fully grown, with a short trunk, long branches, a central root with vertical growth and then developing adventitious roots. They have a reddish-brown bark which is easy to peel off. Bipinnate leaves, with pairs of nodal spines, conical, up to 5 cm long. Bipinnate leaves with petioles 2-14 cm long, rachis may be absent, pinnae 6-24 cm long each with 10-25 pairs of linear leaflets. Inflorescences in clusters, spiciform, 6 to 12 cm long, cylindrical, whitish to yellowish, with 20 to 30 6-8 mm long eHptico-oval seeds. Its fruits are also edible and are used as food for livestock and for humans in the production of alcoholic beverages, flour, water drinks. It is also used for the production of charcoal.

d) Soil characteristics for this species; tolerates sandy soils with salt problems, although sometimes if these soils tend to saline sodic soils they do not tolerate them.

e) Hydrological needs; tolerates drought due to its large root system, exploring wet regions. It is extremely water efficient and has been successfully developed in arid regions.

f) Resistance to low and high temperatures; it does not tolerate frost well, i.e. temperatures below 4 degrees Celsius cause severe damage to its tissues.

g) Recommendations for planting in urban areas; trees in **urban areas**

Urban ornamental and windbreak curtain, also for the use of its wood. Because of their intense bright green leaf appearance, they are suitable for these urban areas due to their excellent foliage and good shading.

h) References:

1. AFPD, 2008. African Flowering Plants Database - Base de Donnees des Plantes a Fleurs D'Afrique

Burkart, A. 1976. A monograph of the genus Prosopis (Leguminosae subfam. Mimosoideae). J. Arnold Arbor. 57(4): 450-525

No.23

a) Common names;

Foreign mesquite or sweet mesquite

b) Scientific name

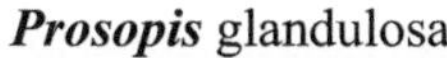

Prosopis glandulosa

23. C. name: Foreign mesquite Scientific name: *Prosopis glandulosa*

c) Description of the species; it is a tree of the leguminous family, native to North America, it is a medium to small tree, with a rounded crown, and stranded, hanging branches with light foliage, pairs of straight thorns, on twigs. It reaches heights of 5 to 9 metres, but can reach up to 14 metres, flowers in the months of March to November with pale, yellow, elongated spikes, and fruits in yellow pods, edible by many species of wild animals, the growth rate of this mesquite is medium. Its aromatic charcoal adds flavour to barbecues, much appreciated in the United States (1). Also native to Mexico, it has been introduced to other countries and is considered an invasive species, i.e. it spreads easily and under very extreme conditions.

d) Soil characteristics for this species; it grows in any type of soil, and can even be observed as an invader in pavements.

e) Hedonic needs; tolerates prolonged drought, and can be found growing naturally in very extreme conditions.

f) Resistance to low and high temperatures; tolerates temperatures too low even below 4 degrees Celsius

g) Recommended for planting in urban areas; in the regions of the countries of origin it is also used as an ornamental, especially for its good appearance at flowering time, however, it is considered to be of medium shade, although it is pleasant to look at, due to its bright colours, especially in spring.

h) References:

Felger, Richard; Mary B. Moser (1985). *People of the desert and sea: ethnobotany of the Seri Indians*. Tucson: University of Arizona Press.

No.24

a) Common names

Wicker

b) Scientific name;

Chilopsis linearis

24. Common name: Wicker Scientific name: *Chilopsis linearis*

c) Description of the species; ***Chilopsis linearis*** described by (1), is a small tree native to southwestern USA and northern Mexico, commonly seen in streams and riverbanks up to 1500 m above sea level, reaching from 1.5 to 8 m in height, having the appearance of other small shrubs, the leaves are linear and curved, reaching 10 to 26 cm in length, and 2 to 4 mm wide, deciduous leaves. The flowers occupy a terminal parncula or raceme, they have 2 to 4 flowers, which open at the same time, the sepals are purple, while the corolla is 2 to 5 cm long.

Lavender to pinkish vanan, the fruit contains a large number of seeds, which are carried by the wind.

d) Soil characteristics for this species; it thrives in poor quality soils of rivers and streams, stony, saline tolerant, moist and well-drained soils. However, it prefers porous, sandy soils.

e) Hyrdic needs; tolerates prolonged drought, and extreme heat.

f) Resistance to low and high temperatures; it is resistant to low temperatures down to minus 15 degrees Celsius. And tolerates high temperatures above 35 degrees centigrade.

g) Recommended for planting in urban areas; in arid and semi-arid regions it is used as an ornamental plant because of its flowering in early spring, the colouring is attractive as the mavon'a of the species is just beginning to leaf out, the shade is not very dense, however, because of its tolerance to low temperatures and resistance to drought it is in great demand for parks and gardens.

h) References:

1. *Chilopsis*". *Useful plants: Linnaeus.* Archived from the original on 23 November 2010. Accessed 4 December 2009.

No.25

a) Common names;

Moorish or Moorish

b) Scientific name;

Morus spp

25. Common name: MulberryScientific name : *Morus spp.*

c) Species description; deciduous trees native to warm and temperate regions of Asia, Africa and North America. They are fast-growing when young, but stop growing at that rate, and rarely exceed 15 metres in height. The leaves are alternate, simple, and sometimes lobed and serrate at the margin. When they are females, which are commonly called blackberries, the blackberries are white, starting with a whitish green colour, and turn white with pinkish edges, when they are ripe until they are purple when they are at the point of harvesting. In the case of the trees that are called blackberries, the inflorescences come out, but after a few days they abort and do not produce fruits, these species are special for ornamental purposes. Blackberries are black with a red colour and are suitable for harvesting and consumption. The unripe fruit and green parts of the plants are known to have a toxic and mildly hallucinogenic saponin. Some types of blackberries, especially the white ones, are economically important, as they are the only source of food for the silkworm, whose pupa cocoons are used to make silk. Because of their high content of total sugars, total acids and vitamins, they are used for juices and juice, as well as for wine making (2).

d) Soil characteristics for this species; they develop in all types of soils, but if these are deep and fertile, their development is exuberant.

e) Water requirements; they have a certain tolerance to drought, however when young and fast growing it is advisable to water them regularly, but once they have grown they can withstand prolonged periods of drought.

f) Resistance to low and high temperatures; as they are deciduous when the cold season arrives, the leaves fall off making them more tolerant to low temperatures in the winter periods, on the other hand, they require good temperatures in summer and autumn for their rapid growth.

g) Recommendations for plantations in urban areas; although they are still found in parks and gardens, blackberries are not very recommendable for ornamental purposes, as when they ripen they fall on the pavements and get dirty and because of their strong colour they remain for a long time, as well as being very appetizing for birds and their excrement is not at all pleasant, on the other hand the blackberry, as it does not produce fruit, is more recommendable for parks and gardens because of its wonderful shade and its good appearance of the colours of the leaves.

h) References:

2. Liu, Xueming et. al. 2004. Quantification and purification of anthocyanins from blackberries with macroporous resins; Journal of Bio Medicine and Biotechnology; 2004:5 326-331.

No.26

a) Common names; Palo Blanco, carob tree or tacus white (5,6)

b) Scientific name;

Prosopis alba

26. Common name: Palo BlancoScientific name: *Prosopis alba*

c) Species description; ***Prosopis alba*** is a tree species from South America, mainly from Argentina extending to Chile. It is a medium-sized tree 9 to 15 metres high and up to 1 metre in trunk diameter. This species is considered to be a nitrogen fixer, being used as a fertilizer for pastures destined for livestock, due to its mineral contribution of this nutrient. The trunk is short and the crown is globular, up to 10 metres in diameter. The branches are thin and often extend to the ground, the bark is thin greyish-brown and woody-grained, with tannic properties. The leaves are pinnate like the carob tree and are very close together. Two to three bipinnate leaves are borne at each node of the stem. Each pinna contains 25 to 40 or more pairs of leaflets, which are glabrous, erect, but somewhat asymmetrical at the base. In winter the tree loses its leaves, but does not defoliate completely. The flower is small, greenish white or yellowish white, hermaphrodite. The pollination is wind or insect pollinated, allogamous, cross-pollinated. The fruit is an indehiscent pod, containing smooth, ellipsoid, laterally compressed seeds, rounded by the pulp, a sweet paste very rich in chlorine, consumed as fodder or made into edible flour for humans. It is also used as charcoal.

d) Soil characteristics for this species; this species prefers loose, well-drained, deep soils and places free from grazing.

e) Water requirements; this tree is perfectly adapted to drought, (xeromorphic), salts and sand, i.e. it is extremely water efficient, produces the most fruits in drought years, has been successful in introducing this species in arid and semi-arid regions.

f) Resistance to low and high temperatures; it does not tolerate frost well, i.e. temperatures below zero degrees are harmful to this species, especially if they are very low and for prolonged periods. It tolerates temperatures above 35 degrees Celsius to a certain extent, however, this causes its growth to slow down drastically.

g) Recommendations for planting in urban areas; It is an ornamental tree used in parks and gardens and in urban areas, its wood is hard, used to make doors or staves in floors, due to its resistance to humidity. It is excellent for outdoor use, it only requires constant pruning to keep it in good shape. It has excellent shade and as it grows it is very comforting.

h) References:

s) Angel Marzocca. 1997 *vademecum of medicinal weeds of Argentina.* Orientacion Grafica Editora, Buenos Aires, Argentina.

t) Enrique L. Ratera and Miguel O. Ratera. 1980. *Plantas de la Flora Argentina empleadas en medicina popular.* Editorial hemisferio sur S.A., Buenos Aires, Argentina.

No.27

a) Commonname

Palo Verde

b) Scientific name;

Parkinsonia spp

27. Common name: Palo Verde Scientific name: *Parkinsonia spp.*

u) Description of the species; described by Charles Linnaeus (2) is a group of plants belonging to the family ***Fabaceae***, native to semi-desert regions of Africa and America. They are shrubs or trees, thorny or not, reaching a height of 5 to 12 m. The leaves are sharply bipinnate with rachises of 2-4 pinnae, very flattened and long, with spinescent stipules and small and numerous opposite leaflets, The leaves are sharply bipinnate with rachises of the 2-4 pinnae, very flattened and long, with spinescent stipules and small and numerous opposite or alternate leaflets. The flowers are hermaphrodite with 5 petals, slightly unequal, and 5 yellow or whitish sepals. There are 10 free stamens, with hairy filaments at their base. The fruit is a tortuous legume, more or less indehiscent, coriaceous, with oblong seeds with endosperm, with a small pical thread. The name palo verde is a reference to its trunk, branches and green pinnae, which are capable of photosynthesis after the fall of the foliage.

v) Soil characteristics for this species; adaptable to soils deep with significant amounts of sand, not very fertile.

w) Hydrological requirements; requires considerable moisture content but has some resistance to drought.

x) Resistance to low and high temperatures; it does not tolerate low temperatures, as very low temperatures and prolonged periods of time can cause death.

y) Recommendations for plantings in urban areas; this species it has important characteristics to be ornamental since it gives good shade and has a pleasant appearance, this for having a fine bearing and also the greenish colouring of its leaves makes it more attractive.

z) References:

2. *Parkinsonia*". *Tropicos.org. Missouri Botanical Garden.* Accessed 7 March 2015.

No.28

a) **Names**

common; Pata de Vaca, ox's foot, ox hoof

b) **Scientific name**;

Bauhinia forficata

28. Common name: Pata de VacaScientific name : *Bauhiniaforficata*

c) **Description of the species**; it is a tree of the fabaceae family, basically distributed in Argentina, and in Mexico in Nuevo Leon and Tamaulipas. It is considered an evergreen tree that can reach up to 7 metres in height. Its flowers are pinkish white, being similar in structure to an orchid. The leaves have a simple alternate disposition, with a lobulated, split margin (resembling the imprint of a fish, hence the name "pezuna de buey" or "pezuna de buey"). Or cow's nipple. The leaf shape is orticulate, with palmate veins. They are perennial, with a blade between 5 and 10 cm long, green in colour, with no change in tonal colour. The fruit is an elongated pod, 15-30 cm long, with a dry, hard, brownish-brown covering.

d) **Soil characteristics for this species**; it grows in deep soils that are not very rich in fertility, it is sensitive to alkaline soils with minor elemental deficiencies.

e) **Water requirements**; tolerates certain degrees of drought, although it is advisable to maintain adequate humidity levels for as long as possible.

f) **Resistance to low and high temperatures**; it does not tolerate very low temperatures, as it always has leaves; however, in the winter period it remains inactive and only when temperatures are very low and for prolonged periods can cause the death of this tree.

g) **Recommendations for planting in urban areas**; this species, from an ornamental point of view, flowers in spring and after two to three years the flowers come out with attractive colours, this being the attraction from an ornamental point of view, they do not shade too much, approximately 60 per cent shade with good growth.

No.29

a) Common names;
Pinabete or Arizona cypress

b) Scientific name;
Cupressus arizonica (2)

29. Common name: Pinabete Scientific name: *Cupressus arizonica*

c) Description of the species; it is a tree of medium size, although in optimal conditions of climate, soil and humidity it grows well. It is a perennial tree that can reach heights from 10 to 25 metres, the trunk can reach up to 1 metre in diameter, the foliage is grey-green or blue-green, in the first years the bark is smooth and when it matures it becomes rough, from where vertical lamellae are detached. Its leaves are grey-green, squamiform or imbricate, with a raised apex,

d) Soil characteristics for this species; it can survive in all soil types, but does well in deep, sandy soils, tolerating high amounts of salts,

e) Water requirements; it is resistant to drought, does not tolerate large amounts of water or being flooded in rainy seasons.

f) Resistance to low and high temperatures; tolerates cold temperatures and frost, however, high temperatures are well tolerated.

g) Recommendations for planting in urban areas; it can be used for reforestation or as windbreaks. From an ornamental point of view, it is suggested to select materials with smaller sizes known as pinabetas, as these have low sizes and also have a good shading condition.

h) References:

2. "*Cupressusarizonica*". *Tropicos.org. Missouri Botanical Garden.* Accessed 31 January 2013

No.30

a) Names

common; Pinguico, Cuban oak or nambimbo

b) Name

scientific;

Ehretia tinifolia

30. Common name: Penguin Scientific name: *Ehretia tinifolia*

c) Description of the species; A phanerogamma species belonging to the Boraginacea, these are trees that reach a height of 15 to 25 m, with essentially glabrous twigs, with persistent leaves, being evergreen plants, with leaf blades of broad leaves from 5 to 14 cm long, eHptic, with leaflets of 5 to 14 mm glabrous, inflorescences 7 to 10 cm long, paniculate, bisexual, bisexual, sessile or short, pedicel uncampanulate, corolla 4 to 6 mm white, style 2 to 4 mm long with truncated stigmas, the fruit is an orange-yellow drupe with endocarp dividing into 2 locular bracts. Pyrenees with raised reticules on the outer surface (4). It grows at altitudes from o to 1900 m above sea level and Mexico in tropical and subtropical areas.

d) Soil characteristics for this species; prefers soils
deep and rich in fertility.

e) Needs 11 id rich; requires significant amounts of water for good growth and development, not tolerant to drought.

f) Resistance to low and high temperatures; although it develops perfectly in tropical and subtropical regions, it has been introduced to temperate zones, however, it does not resist low temperatures, especially below zero degrees Celsius for prolonged periods, which can cause its death; it is adapted to areas with average temperatures of 32 degrees Celsius.

g) Recommendations for plantations in urban areas; this species is considered to be introduced in temperate regions, it has good shade, but does not tolerate low temperatures, from an ornamental point of view, it has a good appearance, however it is considered to be dirty, i.e. it throws away its ripe fruit and stains the pavements, it should be well managed from the point of view of formation pruning to prevent it from reaching flowering and fruit set.

h) References:

4. *"Ehretia tinifolia. Tropicos.org. Missouri Botanical Garden.* Accessed 3 March 2014.

No.31

a) Common names;
Pinon Pine
b) Scientific name;
Pinus cembroides
31. Common name: Pinon Pine Scientific name: *Pinus cembroides*
c) Description of the species; described by (2) species native to Mexico, of the Pinaceae family. It is short in stature, reaching up to 10 metres in height. Its leaves are very small, and its cones measure from 5 to 10 cm. It grows on hillsides and lomerios with scarce slopes at the foot of the mountains, at altitudes of 1350 to 2800 m. It is an evergreen tree. It is an evergreen tree, its crown is round and open in developed trees, and pyramidal in mature trees, with sparse foliage especially in dry places, of medium green colour, sometimes yellowish green, the leaves are in groups of 2 to 3 between 2 and 7 cm in length, they cover the branches abundantly and when falling they leave a scar. The bark is reddish brown or almost black on the outside and is broken into thick layers with small thin scales and deep fissures. The male flowers are in catkins, globose cones, 5 to 6 cm wide, almost stalkless, isolated or in groups of 2 to 5, with large, thick scales, fleshy when green, and brownish green-orange or reddish when the cone matures. The seeds are naked when ripe, and triangular, wingless, 1 cm long, brown or blackish-buff on top and thin at the base. It has a deep root system and monoecious sexuality.
d) Soil characteristics for this species; it grows on poor, shallow soils found on slopes and sides of hills. Lomeria and mountains. Stony or limestone soils,

greyish or black, calcareous with a high gypsum content, thin, well-drained, in lomeria and alluvium in the valleys, with a pH of 4 to 8. And with a pH of 4 to 8, preferring soils with a neutral to alkaline pH.

e) Water requirements; requires constant humidity in early development and once fully developed can tolerate some drought.

f) Resistant to low and high temperatures; grows in dry to temperate-humid, even dry climates, with rainfall up to 800 mm per year, and up to 7 to 8 dry months. It thrives well in dry temperate climates. Temperatures range from 7 to 40 degrees Celsius on average, reaching extreme minimums of minus 7 degrees Celsius.

g) Recommendations for plantations in urban areas; they are species of good growth with little shade, but are important from the ornamental point of view due to their large size, they tolerate prolonged droughts and develop in any type of soil, their fruit being edible, but in regions where they are introduced their fruit does not develop. They are of medium size and have a very resistant root system, which makes them resistant to very adverse conditions.

h) References:

2. *Pinus cembroides*". *Tropicos.org. Missouri Botanical Garden.* Accessed 8 April 2013.

No.32

a) Names

common; Pirul, false pepper tree, Gualeguay or aguaribay.

b) Scientific name;

Schinus molle

32. Common name: Lime treeScientific name : *Schinus molle*

c) Description of the species; a perennial, evergreen tree, belonging to the Anacardiaceae family, native to the central Andes, widely distributed in the Americas, in tropical and subtropical areas, but also grows in temperate areas, feral and sometimes invasive, especially in Mexico (4), tree of small to medium size, 6 to 8 metres high, trunk diameter can reach up to 50 cm, hanging branches, outer bark brown to grey, very rough, exfoliating in long, long plates, erect trichomes, or curved, up to o.1 mm long, whitish, dioecious plants, evergreen or deciduous, alternate leaves, imparipinnate, 2 to 28 cm long, and 11 to 39 opposite to alternate leaflets, narrowly lanceolate. . It has terminal and axillary inflorescences, with leafy bracts, 10 to 25 cm long, globose fruit, 5 to 7 mm in diameter, with slender exocarp, deciduous pink to pinkish red, when ripe, compressed seeds, with flat cotyledons (6).

d) Soil characteristics for this species; this species can develop perfectly well in poor soils with calcareous characteristics, because they do not require much water they are always found in soils that are difficult to manage.

e) Water requirements; it is a drought tolerant species, tolerant of high temperatures, not very tolerant of frost,

f) Low and high temperature resistance tolerates high temperatures and below zero degrees Celsius can cause damage to the plant.

g) Recommendations for planting in urban areas; in urban areas it is considered to be an invasive species, however, it does spread in parks and gardens because of its good regular shade appearance and especially in regions where water is scarce due to its drought resistant nature, however, it is restricted in these temperate areas by the fact that it does not tolerate low temperatures.

h) References:

Ramn'ez-Albores, Jorge Enrique (30 December 2016). "Prediction of the geographic distribution and niche conservation of an invasive tree in Mexico". *Ecosystems* ***25*** (3): 160-163. doi:10.7818/ECOS.2016.25-3.22.

Accessed 7 December 2020.

b) *"Schinus molle. Tropicos.org. Missouri Botanical Garden*. Accessed 16 July 2013.

No.33

a) Common names;

Pirul Chino, Brazilian pepper, turbinto, aroeira, pink or pink pepper.

b) Scientific name;

Schinus terebinthifolius

33. Common name: Chinese lime tree Scientific name: *Schinus terebinthifolius*

c) Description of the species; it is a small tree 7 to 10 metres high, native to subtropical and tropical areas of South America, it has ascending branches, hanging on the tree itself, 10 to 22 cm long. The leaves are alternate, pinnate compound, with 3 to 15 leaflets, these are ovate, lanceolate or etipitate, 3 to 6 cm long, and 2 to 3 cm wide, with finely toothed margins, with acute to rounded apex, The rachis between the leaflets is usually winged, The plant is dioecious, with small white flowers, The fruit is a small, red, spherical drupe, 4 to 5 mm in diameter, in dense clusters of hundreds of drupes. It has an aromatic latex, which causes skin reactions in people sensitive to latex. It is considered to be one of the 100 most harmful invasive alien species in the world (1).

d) Soil characteristics for this species; basically it prefers well-drained soils rich in organic matter, although it can grow in poor, deep soils.

e) Water requirements; it grows in wet and dry conditions, however, in temperate regions where there is not much drop in temperature it can develop perfectly well.

f) Resistant to low and high temperatures; it is not resistant to low temperatures and its habitat prefers tropical and subtropical areas with high rainfall.

g) Recommendations for planting in urban areas; this tree is considered as an ornamental, because it is attractive, especially in frost-free regions, for its dense, decorative foliage and for the red fruits that this species develops. Its dried drupes are sold as pink pepper. An ornamental species that has become a weed in many subtropical regions with moderate to high rainfall.

h) References:

1. Lowe S., Browne M., Boudjelas S., De Poorter M. (2000). *100 of the World's Most Harmful Invasive Alien Species. A selection from the Global Invasive Species Database*. Published by the Invasive Species Specialist Group (ISSG), a specialist group of the Species Survival Commission (SSC) of the World Conservation Union (IUCN). 12pp. First edition, in English, published with issue 12 of Aliens magazine, December 2000. Translated and updated version: November 2004.

No.34

a) Names
common; Willow, willow of the streams
b) Name scientific;
Salix lasiolepis

34. Common name: WillowScientific name : *Salix lasiolepis*

c) Description of the species; it is a species within the genus ***Salix***, native to western and southwestern North America, in the United States and in Mexico in the southern part, it is found in ravines, on the banks of ponds and wetlands (1). It is a large shrub or small deciduous tree that can grow up to 10 metres tall. The shoots are yellowish-brown, dense hairy when the tree is young. The leaves are 3 to 12 cm long, lanceolate, broad, green on the upper side, glaucous green on the underside covered at first with whitish or rusty hairs underneath, which gradually disappear in the summer. The flowers are in yellow catkins, 1 to 7 cm long, produced in early spring (2).

d) Soil characteristics for this species; according to its development characteristics it requires alluvial, deep soils with high moisture content, found in rivers or streams, or in wetlands.

e) Hydric needs; it requires constant humidity due to its development characteristics, this is regularly found in tropical and subtropical regions, however, in temperate zones they develop in rivers and streams of constant water transport.

f) Resistance to low and high temperatures; it does not tolerate low temperatures and its growth and development is more usual in high temperatures and with high ambient humidity contents.

g) Recommendations for planting in urban areas; although its optimum growth is in humid regions in areas where it is used as an ornamental plant, it does grow, but its water requirement is high, that is, it can be used in lakes or streams developed for ornamental landscaping.

h) References:

1 Abrams, L. 1923. Ferns to Birthworts. 1: 1-557. In L. Abrams Ill. Fl. Pacific States. Stanford University Press, Stanford.

2 Correll, D. S. & M. C. Johnston. 1970. Man. Vasc. Pl. Texas i-xv, 1-1881. The University of Texas at Dallas, Richardson.

No.35

a) Common name;
Weeping Willow
b) Scientific name;
Salix babylonica
35. Common name: Weeping willowScientific name : *Salix babylonica*

c) Description of the species; a tree belonging to the ***Salicaceae family***, native to eastern Asia, specifically northern China. Deciduous tree 8 to 12 metres high and can sometimes grow up to 25 metres, with slender, flexible, long hanging branches almost to the ground. Its trunk has fissured bark. Leaves linear lanceolate, 8 to 15 cm long, acuminate, with finely serrate margins, glabrous and glaucous on the underside when adult. Perioles short, pubescent. The inflorescences sprout together with the leaves, have cylindrical catkins, 2 to 5 cm long with pale yellow flowers that reproduce by anemochory (1,3). It grows rapidly, surviving for no more than 50 years. It grows wild in southern Mexico, mainly in the coastal area of Chiapas, Mexico. It is found on the banks of streams and rivers.

d) Soil characteristics for this species; it requires deep soils that retain considerable amounts of water, when these are not, it tends to dry out quickly, i.e. they must be well drained.

e) Water requirements; requires constant water, does not withstand drought, however, in parks and gardens it can regularly thrive on the banks of lakes or streams with constant water.

f) Resistant to low and high temperatures; frost intolerance, i.e. temperatures below zero degrees centigrade, however, they can develop in temperate zones with the appropriate care, as they develop perfectly well in regions with high temperatures.

g) Recommendations for planting in urban areas; shade tree with a good visual appearance, this is due to the drooping of its leaves or branches sometimes down to the ground, its colour is attractive and by taking care of its shape it can be trained to always provide shade in parks and gardens, although it is also widely used as a barrier to prevent soil erosion.

h) References:

2 . Germplasm Resources Information Network: *Salix babylonica* Archived 13 May 2009 at the Wayback Machine.

3 Huxley, A., ed. (1992). *New RHS Dictionary of Gardening*. Macmillan ISBN 0-333-47494-5.

No.36

a) Names
common; Tabachm and
Flamboyan from Madagascar.
b) Name
scientific;
Delonix regia
36. Common name: Tabachm Scientific name: *Delonix regia*
c) Description of the species; it is a tree of the Fabaceae family. It is
One of the most colourful trees in the world, with red and orange flowers and bright green foliage, it can grow up to 12 to 15 metres long. It grows frequently in tropical and subtropical areas. Its natural habitat is the dry deciduous forests, which means that it sheds its leaves when the temperature drops. Its foliage is very dense and it loses its leaves in areas with dry seasons. However, in less rigorous conditions it is evergreen. The flowers are large with four petals up to 8 cm long and a fifth petal called the standard, which is longer and spotted with yellow and white. The ripe pods are dark chestnut-coloured, 60 cm long and 5 cm wide, and the seeds are small and heavy. The seeds are small and weigh about 0.4 grams. The leaves are 30 to 50 cm long and each has 20 to 40 pairs of compound primary leaflets, also called pinnate, and each of these is divided into 10 to 20 pairs of secondary leaflets. In Mexico, being an exotic species, it is distributed practically all over the country, except for Baja California and the central part of the north of the country.
d) Soil characteristics for this species; it can tolerate soils with an excess of salts, deep soils rich in organic matter and which do not become waterlogged.

e) Water needs; due to its developmental characteristics it does not tolerate prolonged droughts and requires constant water for its normal growth and development, and although it can develop in areas with slight drought problems, it is necessary to take care of it in its first years of growth.

f) Resistant to low and high temperatures; thrives well in frost-free climates, preferably requiring tropical or tropical-like climates to survive. Can tolerate short droughts

g) Recommendations for planting in urban areas; it is considered an exotic ornamental plant because of its abundant flowering and brightly coloured leaves. Its ornamental use is common for its utilization, especially in streets and public spaces, it is used as a shade tree because of its wide spreading foliage.

No.37

a) Names

common; Tree-like thunder, tall privet, Japanese privet, kapok, ligustrum, malmadurillo, matahombres and thunder.

b) Scientific name; ***Ligustrum lucidum***

37. Common name: Thunder Tree Scientific name: *Ligustrum lucidum*

c) Description of the species; a plant species of the ***Oleaceae family***, native to the southern half of China, in South America it is considered to be an invasive plant in native forests, displacing the native flora of the area (1). It grows as an evergreen tree 3 to 15 metres high, the leaves are opposite, dark green, 5 to 15 cm long and 3 to 8 cm wide. The fruits are globose berries, 0.6 to 1 cm blackish to bluish, shiny.

d) Soil characteristics for this species; it is not very demanding of soils; it thrives from poor, deep soils to alluvial soils.

e) Hedric requirements; tolerates certain drought conditions, especially when developed as a tree type.

f) Resistance to low and high temperatures; if the growing conditions are good, it tolerates certain conditions of low temperatures, but these should not be for prolonged periods; it tolerates high temperatures, especially in temperate regions where the average temperature exceeds 35 degrees centigrade.

g) Recommendations for planting in urban areas; in urban areas it is considered an attractive tree because of its low water requirement, however, from an ornamental point of view, it can be used as a hedge, but as a tree it can be attractive because of its good shade and good appearance as well as an adequate shading condition.

h) References:

1. "Evergreens in the Sierras Chicas de Cordoba, Argentina. 10 March 2018.

Printed by Books on Demand GmbH, Norderstedt / Germany